Welding for Beginners in Fabrication

The Essentials of the Welding Craft

Roger Scates

Text Copyright © 2018 Roger Scates

All rights reserved. No part of this guide may be reproduced in any form without permission in writing from the publisher except in the case of brief quotations embodied in critical articles or reviews.

Legal & Disclaimer

The information contained in this book and its contents are not designed to replace or take the place of any form of medical or professional advice. The information provided by this book is not meant to replace the need for independent medical, financial, legal or other professional advice or services, as may be required. The content and information in this book have been provided for entertainment purposes only.

The content and information contained in this book have been compiled from sources deemed reliable and are accurate to the best of the Author's knowledge and belief. The Author cannot, however, guarantee its accuracy and validity and cannot be held liable for any errors and/or omissions. When needed, further changes will be periodically made to this book. Where appropriate and/or necessary, you are obligated to consult a professional (including but not limited to your doctor, attorney, financial advisor or such other professional advisor) before using any of the suggested remedies, techniques, or information in this book.

Upon using the contents and information contained in this book, you agree to hold the Author harmless from and against any damages, costs, and expenses, including any legal fees potentially resulting from the application of any of the information provided by this book. This disclaimer applies to any loss, damages or injury caused by the use and application, whether directly or indirectly, of any advice or information presented, whether for breach of contract, tort, negligence, personal injury, criminal intent, or under any other cause of action.

You agree to accept all risks of using the information presented in this book.

You agree that by continuing to read this book, where appropriate and/or necessary that you shall consult a professional (including but not limited to your doctor, attorney, or financial advisor or such other advisor as needed) before using any of the suggested methods, techniques or information in this book.

Contents

I. Introduction 6
II. Getting Started 7
   Workplace Safety 7
      General Health and Safety in Welding 7
      Fire and Explosion Prevention 8
      Fume and Gas Control 9
      Electrical Safety 10
      Safe Use of Welding Gases 10
   Equipment 11
      Welders 12
      Electrodes 16
      Clamps 17
      Angle Grinder 17
      Other Equipment 17
   Safety Gear 18
      Helmet 18
      Gloves 19
      Grounding Clamps 19
      Clothing 19
      Other Safety Products 20
III. Welding Techniques 22
   Stick Welding (SMAW) 22
   MIG Welding (GMAW) 24
   TIG Welding (GTAW) 28
IV. Down to the Wire 34
V. Going Off-Road 36

VI. Conclusion 38

I. Introduction

Everyone is familiar with the concept of welding, and quite a few people have picked up the skill as a hobby. Basically, welding is the process of joining two pieces of metal using heat. The joint of a weld that is well done looks clean and is hard to detect. Welding can be intimidating to some people since you need to work with extreme heat. An accident can be even fatal. That is why you need the right gear in addition to some safety practices if you want to try it out. When you get used to it, the welding process can even be fun.

Welding is a common industrial process, often seen in construction, but it also has some applications in your home as well. With this skill, you can make some quick repairs by yourself, thus saving time and money.

When trying to master a new craft, having the right foundation can drastically reduce the learning curve. This book offers exactly that, a solid foundation on which you can build your welding mastery. The art of welding takes a lot of time and practice, that is why a book cannot contain all the information that will magically turn you into an expert welder. Nothing beats the experience of a true master. However, this book tries to help you start on the right path and give you the right basics that can turn a beginner into an expert. The welding craft is explained as a whole which allows you to see the big picture. With patience, practice and dedicated effort, you too can become a master welder.

II. Getting Started

If you are picking welding up as a profession, then you will need a couple welder certifications and licenses. Basically, you need to follow certain procedures which help determine if you have the skills and abilities to produce an acceptable weld. Most of the time, the test is conducted according to a code. There are many organizations out there that offer such a test. Most of the certifications have an expiration date, but you can always renew or extend your certifications under different requirements.

Before you get started with welding, you need to make some preparations. You need to have certain gear and tools. Welding can be dangerous and can cause severe damage if you are not careful. This chapter will address both the equipment you need, how to maintain it for extended use, as well as certain safety procedures.

Workplace Safety

In order to have a safe working experience, you need to make sure that your workplace is free from any possible hazards. Dangers of fires and explosions are always present when you are welding. Welding spatter and fires from hot metal can cause severe burns without proper protective equipment. The fumes coming from the welding can be toxic. You are also prone to fatal electric shocks. There are also other work hazards you need to be aware of such as the dust, which may accumulate to a dangerous level, noise, vibration and radiation, are some of the most common. Here are some safety procedures you should definitely follow:

General Health and Safety in Welding

Before you start welding, make sure that your workplace is free from any combustible materials. If possible, remove everything that you do not need from your workplace. If you plan to weld in your garage, make sure that no flammable or combustible materials are in the area. You can never tell when a wild spark might fly off and set things alight. If you smell gas, do not light any gas torches or use electric welding equipment. Still, do not rely on your sense of smell entirely. When you have your equipment on, it can be difficult to smell gas.

When you are welding, make sure that your eyes are protected and your skin is covered from the arc flash (we will talk about safety equipment later on). Assume that everything is hot just like how you would assume that every gun is loaded. Keep your gloves on whenever you work and never touch the metal with your bare hands while you are welding.

Fire and Explosion Prevention

First off, prevent the ignition of combustible materials that might be near the welding process. If you are welding a metal wall or partition, or if you are welding near one, it is a good idea to check what is behind it.

When it comes to gas equipment, you should keep all regulators and air hoses free from oil and grease because pressurized oxygen can cause these two to spontaneously combust. Also, avoid getting grease or oil on your hands, gloves, or clothing.

When you weld, use the right gas for the right situation. Never, ever substitute oxygen for compressed air. If you are working in confined spaces, then there is a higher risk of having the fuel gases combining with the air to ignite or explode. Therefore, make sure that the welding gas supply line does not lie in a confined space where leaking is possible. Moreover, have fire extinguishers ready in easily accessible locations and learn how to use them.

You should also watch out for explosion hazards. When you are welding, cutting, brazing, soldering on pipes, tanks, drums, or other vessels, you run the risk of causing an explosion if you are not careful.

Every container that was used to store petrol or other flammable substances is very dangerous to work on. A pinpoint of heat can be enough to trigger an explosion or fire. Other containers that are used to store oil, soap, diesel oil, acids (those that react with metal to produce hydrogen), should also be avoided. If your project involves the use of those containers, make sure you clean them thoroughly. A good way to do it is by steam cleaning those containers and fill them with an inert gas like carbon dioxide or nitrogen. Alternatively, you can fill them with water and leave a small vented space at the point where the repair is to be made (to allow for the expansion of liquid). Just washing them with hot or cold water will not be effective. Blowing air to get rid of the residue will not work as well.

Fume and Gas Control

As you are welding, you will be exposed to gases and fumes, which can be fatal. If you work in an unsafe atmosphere with toxic welding gases and fumes, they can cause discomfort, suffocation, fire and even poisoning. This risk is especially high if you are working in confined spaces.

One way you can control fume and gas is by using a different welding technique. Stick welding produces more fume than TIG welding (this welding technique will be explained later).

You can and should use ventilation. Basically, dilution ventilation takes the fumes and gases close to you and disperses them elsewhere. It is common to see welders placing extraction fans in walls and ceilings, and keeping the doors open (or have a large room) to mitigate the potential damage caused by those dangerous gases and fumes. However, if possible, you should install a local exhaust ventilation system to capture fumes and gases.

Electrical Safety

When you are purchasing or installing welding equipment that uses electricity, ask for advice from the supplier or a qualified specialist. Here are a few basic safety procedures:

- Make sure that your equipment has the correct current capacity.

- Provide an isolating switch

- Do not weld if there is a flammable gas or solvent present

- Choose the right rod holder

- Regularly check the electrical safety of the rod holder or welding hand-piece

When it comes to the electrical safety for the use of welding equipment, water will be your biggest enemy. Be careful when you are welding in places where water may be present. Your hands must be dry before you weld. Of course, due to the high heat, it is very natural for welders to get sweaty. Should that occur, take a break and use a wooden duckboard to insulate yourself. In addition to keeping yourself dry, make sure that your protective clothing is dry as well. Do not work in the rain. Do not twist or knot a lead, bend it sharply, drape it over your body, or tack it to a wall. After you are done using the electrical equipment, disconnect it immediately by pulling on the plug (not the lead).

Safe Use of Welding Gases

Your gas cylinders should be correctly labeled to avoid confusion. Plus, it also saves you time to find the right one. They should also be stored in a ventilated area, and properly secured against falls. The fuel gas cylinders and oxygen cylinders should be stored separately, and the gas cylinders should be kept away from electrical apparatus and other sources of heat. It is worth mentioning that you should examine the gas cylinders regularly to make sure that there are no signs of defects, rust, or leakage. Replace them if necessary. If you have empty cylinders, you should store them separately and mark them as empty. Also, remember to keep their valves closed.

When it comes to safety regarding the equipment connected to the gas cylinders, make sure that the flashback arrestors are fitted to fuel and oxygen bottles. The integrity of the equipment should be inspected regularly. This helps you make sure that there are no signs of defects that might lead to accidents. Preventing these things beforehand can be critical. The oily or greasy substance should not be on cylinders, valves, couplings, regulators, hoses, or any apparatus. It helps if you mark the hoses with colors. Since it is most common, try using red for acetylene (or any other fuel gas), blue for oxygen, and black for inert gas and air. Another thing worth noting is that you should not use copper piping with acetylene.

Finally, the welders should know the correct assembly procedures for attaching equipment to gas cylinders, as well as other procedures such as those to test leaks, to light gas torches and shutting them off. They should also know the signs of a flashback, and how to respond, as well as how to check if any damage has occurred to their equipment and what to do if that happens. You should never crack a fuel gas cylinder valve near a source of ignition. If you want to remove a regulator, close the valve and release the gas from it first. Finally, do not use the oxygen cylinder to dust off your clothing or to "sweeten" the atmosphere.

Equipment

Sometimes, you will need a new piece of equipment for each new project. Normally, the manufacturer or supplier of your equipment should give you detailed instructions about how to properly set it up, its safe usage as well as maintenance. Therefore, you must follow these directions to ensure your own safety and the equipment's longevity. If you are just starting out, there are a few basic types of equipment and protective gear that you will need for every welding project:

Welders

You should look for a long-term investment when you are looking to purchase a welder. Sometimes, it is a good idea to test the waters with a used model. However, you can save a lot of money when you choose the right machine the first time around, not to mention that you do not have to go through all the headaches of switching from one machine to another. It sucks to have your welder fail when you are working on a big project. Still, it is worth noting that even the most expensive welder is not always the best choice for every project. Now you might be inclined to go for the cheapest one on your first project. However, that can also be a mistake since the chance for equipment malfunction can be very high, especially if you are a beginner. That is why you should learn first which type of equipment is best suited for your projects.

You should consider a few things when you are figuring out which welder you should buy. Most important things are the voltage power, AC/DC settings, the duty cycle and the type of projects you will see yourself doing the most often.

The duty cycle is a specification which defines the amount of time, within a 10 minutes period, during which the welder can produce a particular welding current safely. The duty cycle is rated in percentage.

To illustrate, a 150A welder that has a 30% duty cycle and needs to be "rested" or cooled down for 7 minutes after it has been used for 3 minutes of continuous welding. If you fail to comply with the duty cycle limitation, you can stress your welder's power generation system, which makes your equipment fail much earlier than expected. Unfortunately, many welders do not come with an internal protection system to withstand that stress.

Many welding equipment manufacturers will identify a particular welder by its maximum possible power generation. Here, it is important to remember that the number stated is the actual maximum value. Therefore, most of the time, it can only produce that amount of power within a short period of time before it needs to take a break. A welder can function at full power for only 20-30% of its duty power or even less.

For example, if you need 150A to do your work, you should get a 250A welder. That way, you can use it at a lower power, thus guaranteeing a higher duty cycle percentage. In this case, a 250A welder is more than sufficient to get a full 100% duty performance. To put it in simpler terms, your welder needs to go for a lower power if you want to achieve a 100% duty cycle percentage. However, that power level may not be sufficient to work with. Therefore, you need to raise that threshold by buying a welder with much higher power. Longer duty cycles are extremely important when you need to spend more time to weld thicker metals. If you are picking up welding as a hobby, then the duty cycle should not concern you as much. Most likely the projects you will find yourself working on may not require you to work with thick metals. However, if welding is your profession, then the duty cycle is a very important factor to consider when you are looking to purchase a welder.

There is also another thing you need to take into consideration, and that is choosing between Stick, MIG, and TIG welders. All three are different in terms of cost, skill requirement, portability, convenience, and the project itself. While a MIG welder is simple to use, it comes at a steep cost and is not as portable as the Stick welder. On the other hand, the TIG welder gives high-quality, clean welds, but it is very tricky to master. We will look into these three techniques in later on.

When it comes down to fusing alloys and joints, stick welding is your best friend although it is not very efficient. The process uses a consumable stick electrode (hence, the name) that works indoor or outdoor, not to mention that the welding process is very simple. However, there is no shielding gas to protect the weld. Instead, the electrode is coated with a flux that covers the weld and protects it. Therefore, the layer must be chipped away after the weld is done. Stick welding does not need shielding gas, it is a popular and cost-effective method, not to mention it being convenient for the welder. This is because one can switch from one metal workpiece to another easily by changing the filler metal rod to match it.

MIG welders are very popular because they often cost less than both the TIG or Stick welders. To top it all off, MIG welders also have comparable power and features to that of the latter, not to mention that they are very easy to learn and can be used for various projects. Because the filler metal is fed through the MIG welding torch, welders can use both hands to hold the torch in place to achieve a more accurate welding rather than having one hand to add the filler metal. The welding is also much faster because of the wire feeder. The MIG process utilizes an inert gas to shield the weld as well as to keep it free from impurities. Therefore, the MIG welding process is neat and easy to clean up because you do not need to chip anything away, which you normally need to do when you use a Stick welder. MIG welders can be used on a wide variety of materials such as aluminum, and it is also used frequently for automotive works. Still, although MIG welders cost less than the TIG or Stick welders, you still need to buy shielding gas and other materials that might push the cost up.

If you need a high-quality, clean weld, then the TIG welder is an excellent choice for you. The reason being that it is far less likely to distort the metal by using a non-consumable tungsten electrode. Splatter will not be a problem since the TIG welder uses the optimal amount of filler metal from the welding puddle, therefore making TIG welding the highest quality weld in every aspect. However, high-quality welds do not come without training. TIG itself is fairly specialized and you need a great deal of training. If you plan to buy a TIG welder, then you will most likely need to take welding classes. Unlike MIG welding that offers the simplicity and convenience of pointing and shooting, TIG requires the use of a foot pedal to help regulate the welding process. Plus, you need to feed the filler rod which is separated from the torch. Although complicated, TIG welders are a favorite among professionals because of their ability to work on a wide variety of metals. This is thanks to the versatility of the argon gas used during TIG welding. Depending on the project you are working on, you do not need to change the gas since the argon gas is capable of welding any metal of any thickness with TIG welders, not to mention that there is no slag to block the view of the weld puddle.

When it comes to power options, you will need a higher power welder in order to work with thicker metals. Sometimes, you will need special power supply set-ups in order to power higher voltage welders. A welder with a lower voltage in the 100 or so will be insufficient for heavy duty jobs, but at least you can plug and use it from any outlet. If your project requires a welder with the power over 200V, then you will need another power supply set-up, not to mention that it will cost you more to run.

When you need to decide between an AC or DC welder, you need to remember that DC offers a steady rate of energy that leads to a higher temperature and deeper weld penetration. On the other hand, AC welders cost less, but there are only a limited number of electrodes. While DC welders are more expensive, they are still a favorite among many welders because of their wide selection of electrodes and other working advantages such as simple arc striking, better penetration, and improved control. If you know that you will need to work on a wide variety of projects, then it is recommended that you consider buying an AC/DC combination welder.

Electrodes

Electrodes are an essential material in welding. Since there are so many to choose from, many beginners tend to be overwhelmed. Fortunately, this basic guide is here to help you choose the suited for you and your projects.

There are two basic kinds of electrodes: consumable and non-consumable. Stick and MIG welders use consumable electrodes. They supply the filler metal that is used in the weld. TIG welders use non-consumable electrodes which are made of tungsten, so they are not burned up during the welding process. Non-consumable electrodes help the generation of heat with the electrical current, and they are ground to form a flat point. The welding arc is conducted from the point of the electrode for TIG welding.

Each of the three main welding processes require specific electrodes. If you are following the Stick welding process, then you will need a consumable electrode that is melted to create a weld joint. For that, the E6010 electrode is the popular choice. E6011, E6013, and E7018 are also preferred by many welders.

Conversely, a consumable electrode wire is used in MIG welding, which is fed through the welding torch. Although larger projects will require thicker wires, you will most likely need only the wires with the following thickness: .023, .030, .035, and .045.

While consumable electrodes are required in Stick and MIG welding, the processes are different. For MIG welding, the consumable wire electrode is continuously used along with a shielding gas, both of which are fed through the welding gun (although MIG offers flux cored wire as well). In order to lay the weld for Stick welding, the consumable electrode is coated in flux. If you are looking for versatility and portability, then Stick welding is your best choice. If you look for a simple, clean, and functional welding process, then it is recommended that you choose MIG welding.

TIG welders use non-consumable electrodes which come in five options. Most of the TIG welding projects require ceriated and lanthanated electrodes. Unless you have a respirator, you should avoid thoriated electrodes because they produce radiation when used.

Clamps

Clamps help you to keep a weld joint properly lined up. Even a slight movement could cost you the time to redo the weld all over again.

The idea is that the more clamps there are, the better. Experienced welders will testify that they would rather stop their project halfway through just to purchase some more clamps. This highlights the importance of clamps in welding. So, get plenty of these. You can never have too much.

Angle Grinder

For TIG and MIG welding, you will find angle grinders to be very valuable tools. As the name suggests, this portable tool is used to grind down the bumps that are left after welding. It varies from 500 to 2500 watts, depending on the size. You should choose one with enough power to sufficiently clean the metal before you start welding it. If possible, try to get an angle grinder with greater power and durability, because you will need to use this very often.

Other Equipment

While these are not entirely necessary, it can help you with your projects. Here is a small list:

- Permanent markers

- Welding Pliers

- Steel wire brush (a steel brush, not stainless steel or aluminum brush)

- Hacksaw

- Welding table

- Chipping hammer

Depending on the project, you might need more or less than the above. Still, having these ready at all time can be very useful.

Safety Gear

In the welding world, safety is not optional. Metal and extreme heat are two of the most dangerous things you could ever work with. One should always prioritize safety over everything else. Below are some of the must-haves:

Helmet

A good helmet should protect your face from the sparks and your eyes from the light. If you are using the electric arc, then welding will produce ultraviolet light rays that are so bright that they can cause severe damage to your eyes.

Some welders recommend buying an auto-darkening helmet, which frees up their hands. With two hands on the job, it is a lot easier to achieve an accurate weld. Precision is necessary for a good weld. However, these items can be expensive and the darkening effect might be slow. Therefore, some welders prefer wearing some other eye protection instead.

Gloves

It should go without saying that you also need to protect your hands since they are the closest to the heat. The best material for your gloves is the top-grain leather, which is the outer layer of an animal's hide. Welding gloves should have a balance between flexibility and heat protection. Different welding process requires different types of gloves. If you follow the Stick welding process, then you will need stiff, heavy-duty gloves that can withstand high temperatures since that welding method produces the highest heat. Conversely, TIG welding produces the least amount of heat and therefore, the gloves you should use are the ones that are light and flexible.

Many welders favor goatskin leather gloves for TIG and MIG welding. Deerskin gloves are also an equally excellent choice since they have a feature of shaping themselves to the users' hands over time to achieve a very comfortable fit. Top-grain pigskin, elk skin, and cowhide gloves are also good materials for your welding gloves.

Grounding Clamps

Grounding clamps are used in order to protect the user from electric shocks. These are a very crucial safety measure. A good grounding clamp will not only protect the welder but also make it a lot easier to start an arc. The best ones are made out of copper.

If you want them to be effective, make sure that they have a good contact with the workpiece. Some welders even went the extra mile and added a large piece of copper cable to them so they could have better contact.

Clothing

Welding would undoubtedly lead to having sparks fly everywhere, and no one wants their clothing to be on fire while they are working. You should avoid using synthetic shirts since they are not made for this type of work. Wearing this fabric can actually become quite dangerous for you once you start working.

Since the rays from the welding can cause sunburns on exposed skin, you should wear long sleeves shirts. There is a wide variety of clothing to choose from for your welding projects, but what you will wear also depends on the temperature of the work area.

Leather clothing is the safest choice of all, albeit being the warmest and therefore can be quite uncomfortable if you need to work for a prolonged period of time. Many welders often wear leather sleeves, a leather apron, and longer gauntlet gloves, or some sort of welding bib with an open back and long sleeves. Because cotton is not as flammable as synthetic materials, it is wise to wear a cotton shirt underneath your protective gear. It is worth mentioning that you should wear dark clothing to avoid UV rays, and you need your overall to be flameproof, just in case.

Other Safety Products

Your safety needs will vary depending on the kind of project you will find yourself working on. Your worksite or workshop should have at least a ventilation system and a fire extinguisher should things go wrong. Although fume extraction systems are the best thing you can have in your workshop. You can get away with just an open garage door and a simple ventilation system set up in the window. This is acceptable only if welding is just a hobby to you and you want to do it from the comfort of your own home. Still, it is worth stressing that ventilation is a must since certain materials and welding processes can produce toxic fumes.

In addition, having respiratory protection in addition to ventilation will also help, especially when you are welding galvanized metals or when the metal you are welding has cadmium. If that is the case you might need a simple mask that fits under your welding helmet. Eye protection may be worn under a welding helmet for additional protection. If you are using a chipping hammer or powered grinder to prepare a surface or remove slag, then the eyewear will come in handy. Plus, having some sort of hearing protection will help as well. You not only need to work in close proximity to heat, but you also need to deal with the loud noise as you work as well. This can damage your hearing in the long run.

If you are welding at home, then you might want to consider investing in some welding screens so the rays will not harm other people who walk by when you are working. You should also have fire retardant barriers. That way you can prevent sparks from catching on any cloth, cardboard, and (especially) sawdust in the garage. Thys way you can easily avoid unnecessary accidents. Once a fire starts and gets going, it can be very difficult to stop. It's best to avoid these sorts of things at any costs. In addition, having those screens around helps you keep the rest of the place clean if you engage in a messy project.

III. **Welding Techniques**

There are a wide variety of welding techniques you can use to take on any project. However, there are three common welding techniques that you need to be familiar with if you want to become a skilled welder.

Stick Welding (SMAW)

Shielded Metal Arc Welding (SMAW) or Stick welding is among the easiest welding techniques. It can be performed on metals that haven't been pre-cleaned or that are rusty. Meaning that the welder does not need to spend much time on the project using this technique, although it does help if you take your time to clean it beforehand. Moreover, this technique is very effective in welding different alloys and joints. If you are using this technique, then you should know that it is crucial to use the right amount of amperage for the thickness of the metal you are working on. The length of your arc should not be longer than the thickness of the metal itself. Moreover, if you hold the electrode too close to the metal you are welding, then the amperage of your welder will decrease. You should hold the rod perpendicular to the surface with the top of the electrode at an angle of about five to fifteen degrees when you guide it along the surface of the metal. You should also adjust the speed to keep the electrode at the top 1/3 of the weld pool.

For this technique, many welders prefer DC power over AC because a stick weld is a lot easier to start. Moreover, there is less splatter, fewer arc outages and sticking, a cleaner weld, better penetration (up to 10%), a smoother arc and not to mention that you can easily weld in different positions. That does not mean that you should use DC power all the time, of course. If you are looking to stick weld magnetized parts, then AC power is your best bet.

It is worth noting that stick welding is effective only with metals that are thicker than 18 gauges. You also need to change the rod frequently, not to mention that you will need to deal with splatters more than any other welding technique. You will need to do a lot of cleaning with the welds when you are done. To help mitigate the number of splatters, you should choose the right electrode for the job. Like we have mentioned earlier, you will probably only need to use the following: 6010, 6011, 6013, 7018, and 7024.

It is recommended that you purchase a machine with around 225A to 300A. Most of the projects you encounter will most likely require only 200A but having more power would give you more flexibility when you need to take on bigger projects. Should you need to work with metal that requires more than 300A (over 3/8-inch-thick), you just need to make several weld passes. However, just because you can work with thicker metal with an underpowered welder doesn't mean that you should buy a welder with a lower amperage than the range we mentioned. To illustrate, your 155V welder can only handle metal with the thickness of 1/8 inch while your 220V welder can easily handle one with the thickness of 3/8 inch.

Another thing you should keep in mind is the arc length. It depends on the electrode and the application. When you weld, do not hold the electrode too close to the metal. The electrode will produce erratic arcs, not to mention that there will be high crown beads from the decreased voltage. If you hold it too far, however, there will be more splatter (which translates to more cleaning up after the job is done), in addition, you can encounter a low deposition rate and other issues.

When it comes to safety, you should wear proper safety gear at all times. This technique does produce smoke and fumes, so you will need a respiration apparatus or other means of ventilation especially when you work in a confined space. Have your helmet with the correct lens on at all time. Stick welders produce sparks and splatters, that is why you should also put up welding screens, especially if you are working close to other people or with combustibles or flammable materials.

Before you start welding, you need to prepare your gear and equipment. First, find the two 25mm DINSE sockets on your machine. Those are where you will need to connect your electrode holder (+) and the ground clamps (-). Plug in the welder, but do not turn it on yet. Take the electrode holder cable and plug it into the "+" DINSE connector. Then, take the ground clamp and clamp it to your welding table or workpiece. Next, screw the DINSE plug on the opposite end of the cable into the "-" socket to create a DC Reverse Polarity setup.

Next, put on your safety gear. Stick welding can produce a lot of UV light and radiation, so get some eye protection we talked about earlier in the book. Select the appropriate electrode for the job (E-6011 is good if the metal is dirty, rusty, or painted, E-6013 is for thin metal, and E-7018 is for a clean or critical part), put it to the electrode holder and strike the arc by tapping the work or scratching it. Guide the rod carefully down the joint or piece you want to weld. A good way to do it is by moving the arc out of the puddle to allow it to cool down before you give it a second pass. This movement is called "Whip and Pause". After you have finished, pull the electrode away and place the electrode holder (or stinger) on a non-grounded, insulated surface. You will notice the slag, which coats the top of the bead. Use a chipping hammer and give the bead a few taps and rakes. You can do some more cleaning by using the stainless-steel wire brush.

MIG Welding (GMAW)

Gas Metal Arc Welding or MIG welding, requires you to clean the metal before you start working on it. If you want to create a high-standard weld, then you should remove all rusts, paint or solvents that might have found their way onto the metal. Detergent and water, or solvents can be used to remove any grease on the metal if you do not have a grinder at hand. When the metal is properly cleaned, you can start welding. Make sure that, as you are welding, the welding torch does not touch the metal that is being welded. Guide the welder in a weaving design or zigzag pattern to make sure that you weld both sides of the joint. To achieve proper coverage produced by the shielding gas, push the touch instead of pulling it.

MIG stands for (Metal Inert Gas), and this welding technique utilizes electricity to melt or fuse pieces of metal together. MIG is one of the most basic welding styles out there. You use a wire electrode to create a flaming arc that produces heat and creates filler material. As that is happening, an inert gas is released around the weld area to protect it from external contamination.

If welding is your hobby, then you will find that it can be used for many basic household projects like repairing a swing set, a metal fence, or automobile. You do not need a lot of training in order to use this technique since it is very simple, and it produces a relatively clean weld, not to mention its versatility. Since you do not have to change rods or chips and don't have to brush the weld as often, you will have more time working on the weld itself. You do not need to clean up as much and you can weld in any position on various types of metals. This includes stainless steel, mild steel, and aluminum and many more. You can weld quicker, with higher efficiency, because the wire is continuously fed and you can concentrate on the arc control. There are minimal weld defects and there is often little to no slag produced.

Of course, MIG welding does have its own disadvantages. The shielding gas bottles can be annoying when you are welding. You will need to replace the tips and nozzles often, which will cost money. Even if you do not need to spend as much time replacing various parts, as it is in the case of other techniques, you still need to make sure that the metal is properly cleaned. That means, no rust or dirt if you want to have a good weld. Moreover, the penetration of this technique is underwhelming when you are working with thick steel.

When you use the MIG welding techniques, keep in mind that you need to choose the correct welder. MIG welders come in a 1-

Phase, combination of 1 and 3 Phase, and 3 Phase version. The 1-Phase MIG welders normally have a 115 or 230 VAC, and they are used in body shops, on farms and ranches, general repairs, small maintenance, light fabrications, and home garages. The 1- and 3-Phase MIG welders are your go-to welder if you need flexibility when you weld and they have the same application compared to the 1-Phase welder. You can even use it in light manufacturing. The 3-Phase MIG welders are normally used in manufacturing and fabrication. Therefore, you will most likely need only the 1-Phase welder and the combination of 1- and 3-Phase welder for your home projects if you pick up welding as a hobby.

Before you start welding, make sure that you have set everything up for the job. Make sure that the valve to the shielding gas is open and that you have about $20 \text{ft}^3/\text{hr}$ flowing through the regulator. It goes without saying that the welder has to be on and even if this was mentioned before, having a routine check-up in these kinds of situations really helps avoid mistakes and accidents. It is recommended that you create a step by step procedure before you start any welding job. That way, with time, it becomes second nature and you can become much faster and more skilled at this craft.

After you have your welder set up and your piece of metal ready, it is time you start welding. If you are unfamiliar with this, you can practice first. You can do so by trying to run a bead first before you move on to weld metals together. Start by taking a piece of scrap metal and try to weld in a straight line. Rinse and repeat until you feel comfortable with the operation. Since every welder is different, you will need to figure out these settings yourself. If you use too little power, you will have splattered weld that will not be enough to penetrate through the metal and form a proper weld. Too much and you will cut right through the metal entirely because you melted the metal. You will know the right setting when the welds look nice and smooth.

Laying a bead is not really difficult once you get the hang of it. Basically, you draw a zigzag pattern with the welder. Alternatively, you can try to draw little concentric circles from the top of the weld and work your way down. Imagine you trying to "sew" two pieces of metal together (laying a bead is similar to that). Start by laying beads about an inch or two long only. If any weld is too long, then the work will become heated in that area and could become warped. To prevent overheating, just weld a little bit in one spot before moving to the next and come back later. You should weld from the bottom to the top and push the weld with the tip of the gun. However, it might not be comfortable for you. If so, you can work in whatever position or direction that is comfortable, as long as it works. After you are finished, you will see that there are bumps where the filler came in. If you don't really care about the presentation, then your work is finished. If you do care about looks, then you can grind it down and smooth it out.

You can do that by putting the grinding wheel on an angle grinder and use it to smooth down those unsightly bumps. It is worth mentioning here that you should try to make your weld as neat as possible. That way, you do not have too much grinding to do, to polish up your work. Be careful, however, you might accidentally grind through the new weld. Guide the angle grinder around carefully and try not to grind away at a single spot too much. If you see a blue tinge coming from the metal, you are probably pushing the grinder too hard or you are not moving the grinding wheel around enough. Make sure to have protective gear on at all time, even when you are grinding. Small fragments of metal might fly along with the sparks and light. Make sure that you don't have any dangling pieces of clothing because if they get caught on the grinder, you will get sucked in very quickly. Take breaks as needed and stay hydrated.

Even though MIG welding is easy to master, you will still need some practice. Until then, you will run into problems. If the weld starts splattering little balls of metal and turns colors of brown and green, then you do not have enough shielding gas surrounding the weld. Turn up the pressure on the gas and see if it helps. When there is too much metal in your weld pool or the weld is like oatmeal, then you should slow down your wire speed because there is too much of it coming out of the gun. If the welding gun just spits and does not maintain a constant weld, then you might be holding the gun too far away from the weld. The ideal distance is around 1/4 to 1/2 inches from the weld.

There are situations when you weld too close to the material or you build up too much heat at the tip of the wire. This can cause it to actually weld itself to the tip of the welding gun. You will know this when the wire won't come out of the gun anymore, and when there is a little blob at the tip of the gun. First, switch off the welder and unscrew the welding shield cup and the bad welding tip. Slide a new tip into place and screw it on. Replace the welding up and you are good to go.

There are also situations when the wire gets kinked. As a result, it won't go through the hose, let alone the gun, even though you have cleared the tip. You might want to take a look inside the welder, check the spool and the rollers. If the wire gets kinked in there, then you need to replace it through the hose and the gun. To do that, first, unplug the unit and find the kink or jam in the spool. Cut the wire and pull it out from the gun using a pair of pliers or wire cutters. Since the wire is long, you will need to keep pulling for a while. Then, fix the wire and feed it back to the rollers. Some machines have a tension spring you can release in order to insert the wire. Make sure that the wire is seated properly between the rollers and tighten the tension bolt again. Finally, turn the machine on again and depress the trigger. Hold it down until the wire comes out at the tip of the gun again (it can take a while).

TIG Welding (GTAW)

Gas Tungsten Arc Welding (GTAW) or TIG welding utilizes tungsten rods. This type of rods is best suited to create a neat and defined weld that is visually appealing. This is important especially when the weld is in an easily visible area. If you want to have the best quality TIG welds, you must first clean the metal thoroughly. Moreover, you need to select the smallest electrode possible so that when you weld using this technique, you will minimize any possible weld contamination. This welding technique is very sophisticated, and you need to use different arc lengths for different types of metals. However, it is best if you use the shortest arc length possible for any specific types of metal.

Although TIG welding is a bit similar to MIG welding, TIG welding is compatible with a variety of metals. It is an excellent technique you can use if you want to create visually pleasing projects, even art. The metals you can weld include steel, stainless steel, chrome alloy, aluminum, nickel alloys, magnesium, brass, copper, bronze, and even gold. TIG welding has many applications including home projects, repair, automotive welding, and construction.

This technique creates a clean weld of excellent quality, which is important when the appearance is a priority. Splatter or sparks will not be a problem because you only use the needed amount of filler metal necessary in the welding puddle. Moreover, you do not need to deal with the slag that may block your view of the weld puddle. When you are using this technique, you should also consider that there is no need for you to use flux thanks to the Argon gas that is used. Another great thing about this technique is that there is no need to change the gas source for different projects. Argon gas is suited to weld any metal of any thickness when used in the TIG welding technique.

If you have cleaned the metal from any rust, grease, oil, lead, or zinc before you start welding, then there should not be any harmful fumes or smoke that you need to worry about. Similarly, to MIG welding, you can weld in any position with the TIG welding technique. Plus, the latter is safer in most situations. Still, safety is a priority in any welding project. Therefore, you need to wear proper safety gear whenever you weld, regardless of the technique.

For a difficult welding technique such as TIG welding, there can always occur errors, although they are easy to correct if you know how. One such problem is when you burn through your tungsten too quickly. If that is the case, then there may be an inadequate gas flow. Another reason could be that the welder operates on the electrode positive rather than negative. Using the wrong tungsten size for the project can also cause errors, or even using the wrong gas (oxygen and CO_2 instead of Argon gas) can be equally damaging. If the gas flow is the problem, then check the hose, gas valve, and the torch to make sure that they are not kinked or restricted by any means. Plus, check the gas tank to make sure that you still have enough gas to continue welding. For most of the TIG welding operations, the gas flow should be set at about 15 to 20 cfh. The size of the tungsten used for a general-purpose weld is about 3/32 inches at the maximum of 220A.

The tungsten that is normally used comes in five commonly used grades that are color-coded. Green means pure grade, it provides good arc stability and it is also the cheapest. Orange means ceriated tungsten, and it has an easy arc starting, longevity, and good arc stability. Red means thoriated tungsten, and it has a higher current capacity than the ceriated tungsten, not to mention that it has a high resistance to weld pool contamination, however, it is harder to maintain the balled end on AC. Gold means lanthanated tungsten, and it is quite similar to thoriated tungsten. Brown means zirconiated tungsten, and many welders prefer using this over any other type of tungsten. It is excellent for AC welding, and you will always get the best welding result thanks to the quality of the tungsten, although you will need to pay a bit more for it.

Before you start welding using the TIG welding technique, you need to have the right electrode. If you are welding aluminum, then use a pure tungsten rod (green). Once you have selected the right tungsten for the job, grind it to a point if it doesn't have a pointed or rounded tip yet (rounded tip is caused by heat when you weld). For an AC welding, use a balled tip. For a DC welding, use a pointed tip. To insert the electrode, unscrew the back of the electrode holder, put the rod in, and screw it back. Make sure that the electrode remains about 1/4 inches from the sheath.

Next, select the right settings. There are three electricity options: AC, DCEP, and DCEN. Use AC if you are welding aluminum. Use DCEP (or DC, Electrode Positive) to produce a balled tungsten tip, or if you want to stick weld. Use DCEN (or DC, Electrode Negative) if you want to weld steel. The two DC settings differ in the flow direction of the electricity (from the electrode to the metal, or vice versa). You will get a drastic change in the amount of heat the metal absorbs as well as the width and depth of the weld depending on your DC setting. There is also another setting called "cleaning/penetrating". This controls how deep you want your weld to be, with "cleaning" being shallow and "penetrating" giving deeper welds.

Next, you need to choose the gas that is best suited for the project. Most of the time, you will need pure argon or a mixture of argon and carbon dioxide. Argon is for welding aluminum and the mixture is for steel. The gas is important to keep the weld from becoming corroded since it will rust (or oxidize) very quickly at high temperatures.

Next, prepare a metallic area to let the electricity flow through your metal. You can purchase a welding table or just use a large piece of flat metal sheet. If you want your weld to look good, use a wire brush to scrub the surfaces of the metal before you start welding. You can take it a step further by wiping the welding rods with acetone. Of course, cleaning is optional, but you will have a weaker weld, not to mention that it will not look very pretty.

Next, clamp the metals to hold them in place. If possible, spray the welding table down with some anti-spatter to make sure that, should any metal leak off, it will not stick where it lands. That way, you can clean up easily and it helps you to keep the surface flat.

As always, make sure you have the proper protective equipment on before you start welding. You will get sunburns and your eyesight will be affected if you do not take this part seriously.

Therefore, be cautious and have the right gear on. Finally, before you get down to business, check if you can guide the electrode freely.

Start by holding the electrode about an inch away from the metal. If you accidentally touch the metal with the electrode, the molten metal will catch onto the electrode. Stop welding, switch off the electrode, take out the tungsten rod and grind it again. Apply current and heat into the metal by quickly jamming down on the foot pedal. You need to heat the metal quickly and start the weld pool, which is when the metal becomes fluid. Start on an edge and touch the rod in when you see that the pool has been formed. Be careful not to heat the metal for too long since it will warp the longer it is heated. Once the pool is there, go easy on the pedal to be able to control the heat. Remember that welds shrink when they become cooler. You can get your weld off by quite a long distance if the heat is too high for too long. Switch from one side of the metal to another to prevent this. What was previously described is called a "tack" weld. It is basically used to help you hold the metal in place so you can remove the clamps and start working on long bead welds.

After you have finished with the "tack", you can remove all the clamps. You need to start another weld pool, so jam the pedal fast and slow down to regulate the current and maintain an adequate level of temperature. If the metal is starting to burn or melt away, then that means you apply too much current, so slow down. Conversely, if you see that the metal gets flaky but does not look liquid, apply more current.

If you are welding aluminum, then move the electrode to the rod while feeding the rod into the pool. The rod should be to the side of the electrode so that the weld will form or grow. This is called "leading" the electrode.

If you are welding steel, then guide the electrode down the weld line first before the rod "follows" the heat.

TIG welding mostly involves getting the weld pool to form on both pieces of metal at the same time. "Fillet" weld is the easiest type of weld in which you join two pieces of metal at right angles. "Lap" weld involves having them resting flat against each other. "Butt" weld is a bit trickier, which requires them to touch along the edges. The reason why it is tricky is that it is hard to keep the electrode traveling in a straight line along the joint and maintain the weld pool on both of them.

You also need to decide if you want to go for cosmetic or strength welds. Cosmetic welds look more aesthetic even over long lengths. You can create them by making a weld pool, dipping the rod in, and then move to the next point. Cosmetic welds are recommended to weld visible joints such as those on bicycles, where the weld would be exposed all the time.

IV. **Down to the Wire**

An important component for any welding project is the welding wire. It is a slim, metallic rod that is ignited to produce a heated arc to fuse metal together. This is achieved by softening the wire and by hammering or compressing it under an applied heat source. Welding wires are available in a variety of types and classifications. Depending on the job and base metal material, you will need a different welding wire that is appropriate for the application. There are three main types of welding wires. They are solid gas metal arc welding (in short GMAW) wires, composite GMAW (metal-cored) wires, and gas-shielded flux-cored arc welding (or FCAW) wires.

If you want to achieve quality welds, you need to select the right filler metal (welding wire) by knowing its classification. In addition to other equipment and skills needed. The American Welding Society (AWS) classifications for filler metals provides all the information about the wires such as their usability, including the materials they are best suited for and how to use those products to maximize performance. Plus, they also offer insights into the mechanical properties that a filler metal provides.

Basically, the AWS classifications set the standard for filler metals. By learning these you will have a general understanding of what a stick electrode or wire will be able to produce. With the knowledge offered by these classifications, beginner welders are able to become familiar with the product and are given some guidelines on how to operate it. The most important information you can get from the AWS classifications is from so-called designators. They will tell you whether the product is a stick electrode, solid wire, or tubular wire. You will also know about the position the wire should be used, its strength classification, and its chemistry or composition.

For example, E7018-1 H4R is an AWS designation that provides the standards for a common classification of a stick electrode.

E - stands for electrode

70 - describes the tensile strength in KSI

1 - is for the position (1 for flat, horizontal, vertical, or overhead position and 2 is for flat and horizontal position only)

8 - is for the type of coating and current

1 - means that the electrode meets the lower temperature impact requirements

H - means that the electrode meets the requirements of diffusible hydrogen testing

R - means that the electrode meets the requirements of the absorbed moisture test

V. **<u>Going Off-Road</u>**

This part tries to illustrate just a fraction of what off-road welding actually means. We will not go into any details regarding the off-road welding craft, however, you can become more familiar with this subject. Since this is a very complex topic, this part offers a short introduction which can give you an advantage later on, if you wish to learn more about it in the future.

The world of Off-Road and 4x4 vehicles has been blowing up in the past few years. With it comes the need for off-road welding. Even if you are handling a project of building a weekend trail truck or a rock crawler, you know that your ride needs to be able to withstand some abuse along some of the roughest roads out there. When that happens, things tend to break down unless you know what you are doing.

Unlike welding at home, you may have access to a very limited number of tools and gears at your disposal. As such, you will often have a compromised setup that promises no beautiful, professional-looking welds. At least, you can fix things (like the toe-rod) without having to wait for help when you are in the middle of nowhere.

You need to decide what, when, and where you will be using your welder. There is no "master" welder that fits all situations, so select one that suits your needs. There are four general welding techniques in this field: Gas Welding, Arc Welding, Wire Welding, and TIG Welding.

Gas welding uses oxygen and acetylene to, with a help from the striker, produce a flame hot enough to produce weld puddle. Then, add some filler rods, usually mild steel. It is as simple as that.

Arc welding (or Stick welding) still has a place in the off-road world along with MIG and TIG welding. Arc welding is great for heavy fabrication to make strong welds that can take some hits.

Wire welding has two common types of electric wire welders. They are MIG welders and Flux Core welders. Both have their pros and cons, but both are good in fixing cracks and filling gaps.

TIG welding, again, is the most difficult of them all. In the off-road world, it goes beyond beauty. If done correctly, TIG welds can be very strong, which makes the struggles to learn this technique worthwhile.

VI. **Conclusion**

To conclude, welding is a very useful skill to have. With it, you can make certain repairs at home without hiring any professional. That means you can save money and look cool doing it. Plus, you can take on some fun projects like making arts from welding. There are many welding techniques out there, but the most common and useful are by far Stick Welding, MIG welding, and TIG welding. With those three techniques, there are no projects that you could not handle.

It is worth mentioning again and again that safety is not optional. Although to some people, welding might look like fun, it is still a relatively dangerous activity. When you are welding, make sure that you have the best tool for the job so that you can save yourself from wasting time and money only to achieve sub-par welds. Moreover, make sure you have the proper protective gear on at all time when you weld. Tidy up the workplace, and make sure that it is a safe environment to work in. If possible, have only the tools you need in the same room. Be extremely careful around flammable and combustible materials. It can get tedious to clean everything up prior and after every welding session, but it is always better to be safe than sorry.

With patience and practice, there is nothing you cannot learn. There is a lot more to cover if you intend to become a master welder. Hopefully, this book offered you a solid baseline on which you can build your welding expertise. All that remains is for you to start taking action. Now go out there so you can finally let the spark fly, literally.

Made in United States
Orlando, FL
09 August 2022